Sunrises in Motion

A Nature Photo Flip Book Where Motivation Meets Mindfulness

Tim S. Keane

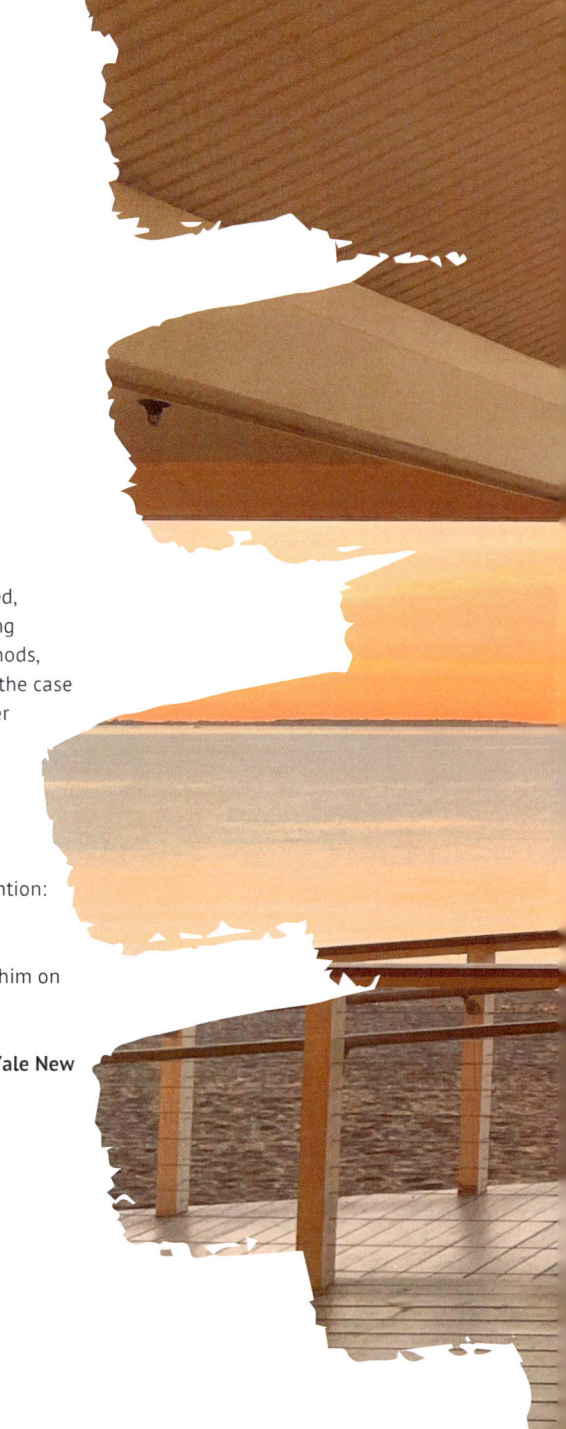

Copyright © 2023 by Timothy S. Keane

All rights reserved. No part of this publication may be reproduced, distributed, or transmitted in any form or by any means, including photocopying, recording, or other electronic or mechanical methods, without the prior written permission of the publisher, except in the case of brief quotations embodied in critical reviews and certain other noncommercial uses permitted by copyright law.

Hardcover ISBN: 979-8-218-28496-1
Paperback ISBN: 979-8-862-26368-8

For permission requests, write to the publisher, addressed "Attention: Permissions Coordinator," at: TSKeane12@gmail.com

Connect with Tim on LinkedIn by searching his name, or follow him on Twitter and Instagram at @GritBalanceMind.

A large portion of all proceeds go to the Joseph Keane Fund at Yale New Haven Health hospitals.

Printed in the United States of America
First Printing: September 2023
Second Edition: March 2024

For Jess and our children...

INTRODUCTION

"REST BUT NEVER QUIT. EVEN THE SUN HAS A SINKING SPELL EACH EVENING. BUT IT ALWAYS RISES THE NEXT MORNING." – MUHAMMAD ALI

Since the beginning of time, the appeal of early morning beach sunrises has captivated many. It initially offered a great way to start the day - a moment for quiet thoughts, joined by a cold brew coffee, while watching an array of colors light up the sky as the sun rises. During the pandemic, an idea emerged: what would it look like, and what would I learn, from capturing an entire year's worth of sunrise photos from the exact same location? So, for the span of a year, a sunrise was captured at regular intervals. This book is a culmination of the project, which is essentially a visual diary of a year's worth of sunrises captured from the same vantage point. But it's not just about the pictures. It's also about the stories, the reflections, and inspiration that each sunrise brings.

A UNIQUE FLIP BOOK EXPERIENCE

First and foremost, this book is designed to be a visual journey. Once I started catching weekly sunrises, I couldn't help but draw parallels to the anticipation of a game's kick-off, or the intro music as a band walks onto the stage. Each dawn brings unpredictability—a "What will the sky do today?" vibe. Will it bring calm gray, pop with vibrant orange, or show off the big sky blues? Sometimes, the sky looks amazing and you can't help but say "wow," especially right before the sun shows up. But those colors are fleeting as they quickly vanish as the sun climbs over the horizon.

MORE THAN JUST PICTURES

There's also a secondary layer to this book as beneath the visual experience lies a collection of perspectives and metaphors related to the sun. The mix of sunrise imagery and thought-provoking quotes adds a layer of depth that goes beyond a simple, one-time flip-book experience.

One morning in particular, during the intense early days of the pandemic's peak, I spoke with an elderly man. While looking at the spectacular pink and orange sky lighting up as the sun came over the horizon, he said, "It's going to be okay." That was all that was said, but it was a great reminder of the importance of perspective.

This moment led me to curating insights from athletes, scientists, and thought leaders to give you a new perspective on the meaning and influence of the sun in our lives. Here, you'll find motivational quotes and scientific tidbits, designed to inspire and educate.

Among them stands NBA Hall of Famer, Kobe Bryant, a basketball player known for his fearless, relentless, and indomitable spirit. Yet, Kobe's narrative transcends the court, as he strived to balance being fiercely motivated with a mindfulness practice in the early sunrise hour. He revealed his approach, an anchored morning routine that allowed him to face the day's challenges with a balanced and focused determination. Speaking of his sunrise mindfulness practice,

"I DO IT FOR ABOUT 10-15 MINUTES. I THINK IT'S IMPORTANT BECAUSE IT SETS ME UP FOR THE REST OF THE DAY. IT HELPS ME. IT'S LIKE HAVING AN ANCHOR. IF I DON'T DO IT, I FEEL LIKE I'M CONSTANTLY CHASING THE DAY, AS OPPOSED TO BEING ABLE TO CONTROL AND DICTATE THE DAY."

Kobe worked at the delicate balance that fuels both motivation and the art of being in the moment.

EARLY RISERS

On any given day, you'll find a few early risers taking in the sunrise at the beach. I've seen people of all ages, from kids with their parents to the elderly. What is it that draws people to a sunrise? Besides the ever-changing sky, there is always something to see.

A small handful of people are typically there. Some are part of the polar plunge group—older guys in their sixties who do their best to fight off aging with cold-water exposure by swimming year-round. You'll see people walking the beach; sometimes there are photo shoots with a full-scale photographer, and sometimes there are teenagers and college students taking their sunrise selfies. I've also caught extraordinary events, such as a rare solar eclipse, which presented an eerie, eclipsed sun in a vibrant orange sky.

PHOTOGRAPHY

As you flip through the following pages, you'll notice not just the movement of the sun, but also the changing tides and shifting sands, reminding us that while some things remain constant, change is inevitable. You'll see some spectacular sunrises as well as some gray-sky sunrises as the seasons change. Upon a second viewing, you may notice another constant: the change in the ocean tide. The horizon line (where the sun meets the sky) remains in the same spot but may deceptively appear to be moving due to the tide changes shown on the beach.

The photos were captured weekly over the course of twelve months from June 2021 to June 2022. Sunrise times ranged from as early as 5:18 a.m. in June to as late as 7:16 a.m. in January. During the winter months, the real feel temperatures dropped as low as -15 degrees.

A NEW PERSPECTIVE

Besides a sunrise, can you think of any other daily event that gives us so much?

The sunrise provides everything from warmth, energy, and weather changes, to motivation, inspiration, discipline, mindfulness, and even sparking curiosity about science.

Let's start with the basic science of the sun. Many people view the sun strictly for the utility it provides that benefits them. We all know the sun offers a lot of tangible uses: heat, energy, weather, and light. This allows for human life on earth.

And then we see others who view the sun as a source of motivation, inspiration, discipline, mindfulness, and science. It's clear that the sun provides more than just its warmth to many. The common thread in most of these perspectives is positivity, and how the sun can serve as a mirror reflecting life itself. Let's break it down into different categories of sunrise perspectives:

MOTIVATION

Muhammad Ali is often voted the GOAT (Greatest of All Time) among athletes, and he provides one of the best motivational quotes incorporating the sun. Ali views the sun similarly to a boxer in that it may go down, but it never quits—it always gets up. Many athletes often reference the sun as a metaphor for motivation. World-class athletes like Venus Williams and Kobe Bryant have all spoken metaphorically about the sun, as if it provides them with the extra fuel to push themselves.

Entertainer Kevin Hart has talked about overcoming challenges in a single-parent household in Philadelphia. He has shared his growth mindset in multiple interviews, reminding us that not every day is beautiful but encouraging us to persevere and stay motivated because:

> "ALL STORMS EVENTUALLY PASS AND WHEN THEY DO, THE SUN SHINES BRIGHT..."

Entrepreneur Gary Vaynerchuk, commonly known as Gary V, often uses the sun emoji when delivering written constructive feedback. This serves to remind the recipient that the feedback, although tough to hear, comes with the warmth of someone who believes in them.

These insights encourage us to meet life head-on and give it our best shot, no matter what the challenge is.

INSPIRATION

There is no shortage of inspirational sunrise pictures and quotes, as many people view the sunrise through this lens. One of the most revered First Ladies, Eleanor Roosevelt, so poignantly said, "With the new day comes new strength and new thoughts." Keep in mind that this was a woman married to a polio-stricken husband, FDR, who was in the White House during World War II, dealt with natural disasters like the Dust Bowl that wiped out much of the Midwest, and led a country still recovering from the Great Depression. That quote provides some amazing perspective.

Abraham Lincoln led the country through a violent Civil War, losing many of his country's troops and some of his closest friends. He grew up in poverty and lived alone with his sister for seven months when they were 9 and 12 years old. This happened after his mother died from typhoid, and his father left to find another wife. He also suffered several election defeats, experienced bouts of serious depression, and lived through the deaths of his first fiancée and his eleven-year-old son while in the White House. Through all of this, Lincoln persevered and is often viewed as one of the greatest leaders of our country. A plaque at a historical site in Illinois captures a line he shared in a speech:

> "BEHIND THE CLOUD, THE SUN IS STILL SHINING."

Jocko Willink is revered as one of the most visible and inspiring former Navy SEALs. He has lost peer Navy SEALs in combat during the Iraq War while serving his country. No matter how bad things are, Jocko reminds us, "There will be dark times. But the sun will come up."

Think about what each of the above people experienced in their careers—including the struggle and deaths of loved ones—but continued to move forward while using the sunrise as an inspirational metaphor.

MINDFULNESS

The key takeaway for those viewing the sun through a mindfulness lens focuses on the importance of being present and self-aware.

Dwayne Johnson, aka The Rock, views his sun and sunshine tattoos as enduring symbols of overcoming struggle and recognizing gratitude. He's had a fascinating career in which his early life was often troubled by run-ins with the law. He then had a football career derailed by injuries and reached a point where he only had seven dollars to his name—hence the name of his business venture, "Seven Bucks Productions." He persevered and eventually transitioned from wrestling to Hollywood.

Entrepreneur Tim Ferriss has put a focus on mindfulness, and his dedication to self-awareness is evident as he speaks of the joy he finds in simply walking in flip-flops while bouncing a handball under the sunshine.
These insights remind us that life doesn't happen in a straight line, but we should embrace and find joy in the simple things.

DISCIPLINE

Some people view the consistent sunrise as a metaphor for personal self-discipline. What better example for discipline is there than retired Navy SEAL and author of Discipline Equals Freedom, Jocko Willink? Jocko uses the spectacular visual of a sunrise as his reward for completing his daily workout before the sun rises.

Entertainer Kevin Hart spoke passionately about the metaphor of sun discipline in an interview. He noted that the sun continues to rise each day, regardless of how it's feeling or what else is going on around it, and consistently does its job of lighting up the globe with its warmth.

Neuroscientist Kelly McGonigal, Ph.D., encourages us to develop the discipline to rise each morning to experience the quiet stillness of the morning—an ideal time for deep thinking.

> "RISING A BIT EARLY CAN ALLOW YOU TO EXPERIENCE INNER STILLNESS AND OFFER YOUR ENERGY TO A GREATER INTENTION FOR YOUR DAY."

These insights encourage us to view the sun as a metaphor for self-discipline and to consistently rise and embrace the day.

SCIENCE

Astrophysicist Neil deGrasse Tyson always does an excellent job of breaking down research, and a few of his insights are shared in this book. For example, he explains that it takes sunlight over eight minutes to travel from the sun to our eyes. In other words, if the sun were to disappear, as it does at sunset, we'd continue seeing it for another eight minutes. That's a fun fact.

Another angle of science and the sun focuses on how it impacts the human body. Stanford neuroscientist and podcaster Andrew Huberman, Ph.D., often speaks about the importance of getting into the sunlight as soon as you wake up. Research shows that sunlight impacts many aspects of human physiology, including your energy level, mood, and circadian rhythm, which helps improve your sleep patterns.

These insights, particularly from Andrew Huberman, have genuinely opened my eyes to the benefits of getting into the sun whenever possible, as it provides a performance edge.

LET'S FLIP THE PAGES AND
EXPAND YOUR HORIZONS.

January week 1

"Behind the cloud, the sun is still shining."

**Abraham Lincoln,
16th US President**

January week 2

Due to refraction of sunlight, *"When you see a sunrise, the sun is still below the horizon and has not yet risen. You're getting a few extra minutes of sunrise light."*

Neil deGrasse Tyson, PhD, Astrophysicist

January week 3

"By looking at sunlight through a window, it's 50 times less effective than if that window were to be open — mostly because those windows filter out a lot of the wavelengths of blue light that are essential for stimulating the eyes and this wake-up signal."

**Andrew Huberman PhD,
Stanford University
Neuroscientist**

January week 4

"We have always held to the hope, the belief, the conviction, that there is a better life, a better world, beyond the horizon."

Franklin D. Roosevelt, 32nd President of the United States

February week 1

"Sun doesn't stop for nobody, man. The sun doesn't stop. The sun is going to be up in the morning, regardless. Regardless of how I feel and how depressed I am, the sun is going to shine in the morning, and at nighttime, the moon is going to be there. And you're going to look up; these days are going to keep going by. So are you going to let the days go by and look up to find you've wasted a year doing... what? Or do you just pick it up, figure it out, admit you've made some mistakes, and move on? Life goes on. I'm a 'life goes on' type of guy." – Paraphrased for content.

Kevin Hart, Entertainer

February week 2

"Remember to look up at the stars and not down at your feet. Try to make sense of what you see and wonder about what makes the universe exist. Be curious. And however difficult life may seem, there is always something you can do and succeed at. It matters that you don't just give up."

Stephen Hawking, Physicist, Cosmologist and Author

February week 3

"I've always loved sunrises because I think there is so much hope in the morning and so much that happens in those moments when you see the world waking up. I feel like it's a time to reflect and think. I love to gather as many people that are willing to come with me and watch the day break together. It always hits me the same way. It's like wow – how small we are in this big beautiful world – and how grateful I am to have this opportunity."

**Lindsay Czarniak,
Podcast Host, ESPN,
Fox Sports**

February week 4

He prefers speaking with the person, but when constructive feedback can only be written, he uses a sun to show his feedback comes with warmth: *"The misinterpretation of the written word digitally is a monster. People will consume it based on their framework. When I do [send a Slack] and I have to give any level of direct [feedback], I'm coming with a heart emoji and a sun right behind it."*

**Gary Vaynerchuk,
Entrepreneur**

February week 5

"You create your own universe as you go along. The stronger your imagination, the more variegated your universe. When you leave off dreaming, the universe ceases to exist."

Winston Churchill, Former UK Prime Minister

March week 1

"If the sun comes up, I have a chance."

Venus Williams, Tennis Pro

March week 2

"Walking in the sun w flip flops bouncing a yellow handball. Does it get any better? Lovely."

**Tim Ferriss
Entrepreneur,
Podcast Host, Author**

March week 3

"How far away is the sun in light travel time? it takes 500 seconds for the light to reach earth from the sun so 500 seconds if you do the math it's 8 minutes and 20 seconds."

Neil deGrasse Tyson, PhD, Astrophysicist

"The universe is change. Life is opinion."

**Marcus Aurelius,
Roman Emperor
during 180 AD, Stoic
Philosopher**

April week 1

"All storms eventually pass and when they do the sun shines bright as hell....Stay motivated & Stay focused....We are all great damn it!!!"

Kevin Hart, Entertainer

April week 2

If the sun disappeared *"from the center of our solar system we would continue to see the sun and feel the sun until 8 minutes and 20 seconds passed."*

Neil deGrasse Tyson, PhD, Astrophysicist

April week 3

"The yellow-blue contrast present at sunrise (& orange-blue at sunset) = the optimal stimulus for setting your circadian rhythm for quality sleep and daytime mood, focus & alertness. Even w/cloud cover when it's not perceptible it's still there & is more effective than any artificial light."

Andrew Huberman PhD, Stanford University Neuroscientist

April week 4

"The universe doesn't allow perfection."

Stephen Hawking, Physicist, Cosmologist and Author

"The task is simple: Do Work. Reward: Sunrise."

Jocko Willink, Ex-Navy SEAL, Author

May week 2

"Each moment of my life I was dreaming of how great I could be, and continued working hard. Each time I closed my eyes I could see me shining bright like a sun."

Kobe Bryant, NBA Hall of Fame Basketball Player

May week 3

Reflecting on his sun and sunshine artwork as they represent: *"the great struggle and overcoming that struggle and being appreciative of my success"*

Dwayne Johnson, "The Rock"

May week 4

"Wow, where did the sun come from? Was it sleeping?"

6-Year-Old Child Watching the Sun Rise Above the Horizon

June week 1

"Sunrise walks teach me the calm and peace of the sun, moon, and stars. When you take a moment to pause, it can provide you with answers."

Sunrise Walker

June week 2

"TODAY is a new day, a new beginning full of POSSIBILITY. Today is more precious than we can even comprehend!"

**Tony Robbins,
Entrepreneur**

June week 3

*"It's a beautiful day,
Don't let it get away"*

Bono, Lead Singer of U2

June week 4

"I like how watching a sunrise makes people happy. It represents new beginnings, as this day has never happened before."

Teenage Girl

July week 1

"It doesn't matter whether you are a Lion or a Gazelle... when the sun comes up, you'd better be running."

Eric Thomas, Author, Keynote Speaker, Coach

July week 2

"Humanity will not be cast down. We are going on—swinging bravely forward along the grand high road—and already behind the distant mountains is the promise of the sun."

Winston Churchill, Former UK Prime Minister

July week 3

"Greatness and nearsightedness are incompatible. Meaningful achievement depends on lifting one's sights and pushing toward the horizon."

Daniel H. Pink, Author, Businessman

July week 4

"Each sunrise is beautiful and peaceful, which brings a new opportunity to consistently pause before attacking each day."

Sunrise Jogger

July week 5

"It's peaceful and grounding. The sunrise is always a little different, but always special."

Sunrise Beach Runner

August week 1

"When I watch the sunrise over the beach, I often think of my dad as he loved this when he was around."

Woman, Mother, and Wife at Sunrise with Her Coffee

August week 2

"If you wake up before the sunrise, try to get some sunshine in your eyes once it comes out. If you miss that morning sun for more than a few days and you view evening sunlight you delay your circadian clock which makes you fall asleep later, wake later etc. Get morning light."

**Andrew Huberman PhD,
Stanford University
Neuroscientist**

August week 3

"A human being is a part of the whole, called by us "Universe," a part limited in time and space. He experiences himself, his thoughts and feelings as something separate from the rest—a kind of optical delusion of his consciousness. The striving to free oneself from this delusion is the one issue of true religion. Not to nourish it but to try to overcome it is the way to reach the attainable measure of peace of mind."

Albert Einstein, Physicist

August week 4

"We accept the voice in our head as the source of all truth. But all of it is malleable, and every day is new. Memory and identity are burdens from the past preventing us from living freely in the present."

**Naval Ravikant,
Entrepreneur, Investor**

September week 1

"At dawn, when you have trouble getting out of bed, tell yourself: 'I have to go to work—as a human being. What do I have to complain of, if I'm going to do what I was born for—the things I was brought into the world to do? Or is this what I was created for? To huddle under the blankets and stay warm?"

Marcus Aurelius, Roman Emperor during 180 AD, Stoic Philosopher

September week 2

"Rest but never quit. Even the sun has a sinking spell each evening. But it always rises the next morning."

Muhammad Ali, Boxer

September week 3

"See the stars, the moon and the sun, how they move in silence... We need silence to be able to touch souls. Silence gives us a new perspective."

Mother Teresa, Nun and Founder of the Missionaries of Charity

September week 4

"Showing up is the key. The sun shows up every day. It strives to provide warmth while lighting up a blue sky. But it gets challenged every day by clouds, storms, hurricanes and more. But the sun doesn't give up. It perseveres and shows up every day. It rises again. And again. And again. Will you?"

Sunrise Runner

October week 1

"With the new day comes new strength and new thoughts."

**Eleanor Roosevelt,
Former First Lady**

October week 2

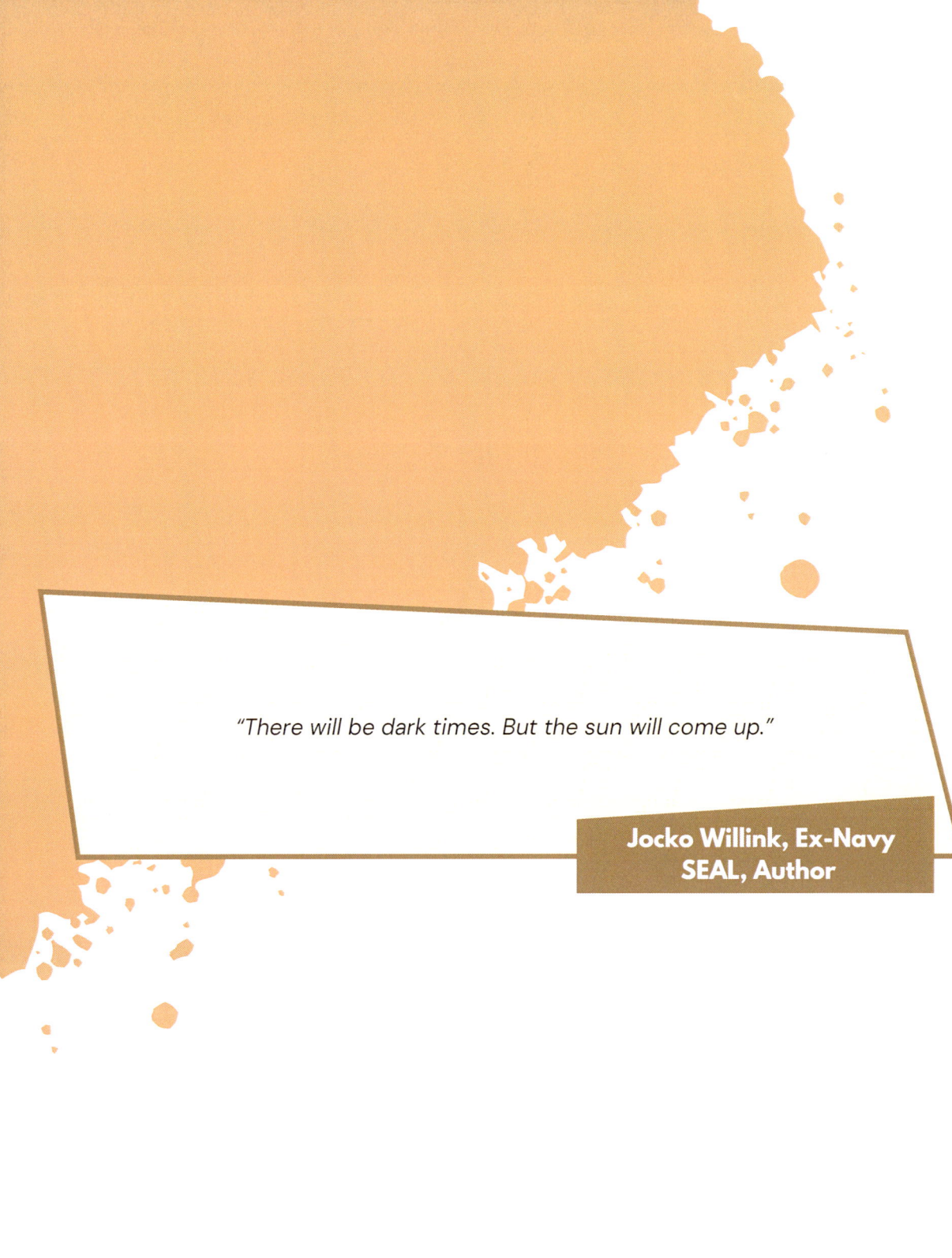

"There will be dark times. But the sun will come up."

Jocko Willink, Ex-Navy SEAL, Author

October week 3

"How many earths would fit in the sun? About one million earths!"

**Neil deGrasse Tyson, PhD,
Astrophysicist**

October week 4

"Some things just go better together, And probably always will, Like a cup of coffee and a sunrise"

Luke Combs, Musician

October week 5

"As long as this exists, this sunshine, this cloudless sky, and as long as I can enjoy it, how can I be sad?"

Anne Frank, 14 year old World War II Diarist

November week 1

"For most of us, early morning is one time of the day we can be alone, without demands and distractions. Rising a bit early can allow you to experience inner stillness and offer your energy to a greater intention for your day."

**Kelly McGonigal, PhD,
TedTalk Speaker,
Psychology Professor at
Stanford University**

November week 2

"Every sunrise is a reminder that life can begin anew, no matter what yesterday held"

Businessman and Scuba Diver

November week 3

Tides rise and fall in the ocean due to the gravitational pull of the sun and the moon. The moon has a bigger influence because it's closer, but the sun's gravity also affects tide changes.

National Geographic

November week 4

*"The sunrise is better
if you already paid for it."*

**Jocko Willink,
Ex-Navy SEAL,
Author**

November week 5

"And sometimes is seen a strange spot in the sky. A human being that was given to fly..."

Eddie Vedder, Musician

December week 1

Articulating his thoughts on nature, human progress and development: *"Sunrise and sunset turned the sky to an unearthly flame of many colors above the vast water. It all seemed the embodiment of loneliness and wild majesty. Yet everywhere man was conquering the loneliness and wrestling the majesty to his own uses."*

Teddy Roosevelt, 26th President of the United States

December week 2

"The sun brings wisdom. And the energy is possibilities."

Husband and Father of Two Boys

December week 3

"When I make it up to the beach for sunrise, I feel like it's Christmas. Will I see the sun? What colors will I see? Every sunrise is completely unpredictable and magical in its own way. Nothing else exists in that moment."

Woman Taking Photos

December week 4

"Even during a quiet night at home, we're spending time thinking of our list of improvements we need to work on. There may be a beautiful sunset, but instead of taking it in, we're busy taking a picture of it. We are not present, and so we miss out. On life. On being our best."

**Ryan Holiday,
Best-selling Author**

REFERENCES

Muhammad Ali, Boxer

- "Rest but never quit. Even the sun has a sinking spell each evening. But it always rises the next morning." Source: Parade Magazine, March 30, 2023 https://parade.com/1401735/michelle-parkerton/sunrise-quotes/

Marcus Aurelius, Roman Emperor during 180 AD, Stoic Philosopher

- "At dawn, when you have trouble getting out of bed, tell yourself: 'I have to go to work—as a human being. What do I have to complain of, if I'm going to do what I was born for—the things I was brought into the world to do? Or is this what I was created for? To huddle under the blankets and stay warm?'" Source: Book Meditations, Chapter Book 5, as translated by Gregory Hayes.

- "The universe is change. Life is opinion." Source: Book Meditations, Chapter Book 4, as shared by Ryan Holiday in Daily Stoic

Bono, Lead Singer of U2

- "It's a beautiful day, Don't let it get away." Source: U2 song, Beautiful Day, https://www.u2.com/lyrics/22

Kobe Bryant, NBA Hall of Fame Basketball Player

- "Each moment of my life I was dreaming of how great I could be, and continued working hard. Each time I closed my eyes I could see me shining bright like a sun." Source: Basketball Mindset, November 14, 2022. https://www.basketballmindsettraining.com/blog/mamba-mentality-quotes

- "I meditate every day. Every day. I do it in the mornings. I do it for about 10-15 minutes. I think it's important because it sets me up for the rest of the day. It helps me. It's like having an anchor. If I don't do it, I feel like I'm constantly chasing the day, as opposed to being able to control and dictate the day. Not that you're calling the shots on what comes forward but the fact that I am set for whatever may come my way. I have a calmness about whatever comes my way. A poise. And that comes from starting my morning off with meditation" Source: Interview video, https://www.youtube.com/watch?reload=9&v=E78y66GEPvs&feature=youtu.be

Winston Churchill, Former UK Prime Minister

- "Humanity will not be cast down. We are going on—swinging bravely forward along the grand high road—and already behind the distant mountains is the promise of the sun." Source: International Churchill Society, June 29, 2013, https://winstonchurchill.org/publications/finest-hour/finest-hour-135/behind-the-distant-mountains-is-the-promise-of-the-sun/

- "You create your own universe as you go along. The stronger your imagination, the more variegated your universe. When you leave off dreaming, the universe ceases to exist." Source: My Early Life, My Early Life, Thornton Butterworth. 1930. https://winstonchurchill.org/

Luke Combs, Musician

- "Some things just go better together And probably always will Like a cup of coffee and a sunrise", Source: Luke Combs' website, Better Together lyrics, https://www.lukecombs.com/track/better-together/

Lindsay Czarniak, Podcast Host, ESPN, Fox Sports

- "I've always loved sunrises because I think there is so much hope in the morning and so much that happens in those moments when you see the world waking up. I feel like it's a time to reflect and think. I love to gather as many people that are willing to come with me and watch the day break together. It always hits me the same way. It's like wow - how small we are in this big beautiful world - and how grateful I am to have this opportunity." Source: Lindsay Czarniak's Instagram, July 21, 2023 https://www.instagram.com/reel/Cu9kd8AMyd6/?utm_source=ig_web_copy_link&igshid=MzRlODBiNWFlZA==

Albert Einstein, Physicist

- "A human being is a part of the whole, called by us "Universe," a part limited in time and space. He experiences himself, his thoughts and feelings as something separate from the rest—a kind of optical delusion of his consciousness. The striving to free oneself from this delusion is the one issue of true religion. Not to nourish it but to try to overcome it is the way to reach the attainable measure of peace of mind." Source: The Ultimate Quotable Einstein, Alice Calaprice. 1950

Tim Ferriss, Entrepreneur, Podcast Host, Author

- "Walking in the sun with flip flops bouncing a yellow handball. Does it get any better? Lovely." Source: Tim Ferriss Twitter/X, October 18, 2008 https://twitter.com/tferriss/status/965643639

Anne Frank, 14-year-old World War II Diarist

- "As long as this exists, this sunshine, this cloudless sky, and as long as I can enjoy it, how can I be sad?" Source: Book, Stillness is the Key by Ryan Holiday, 2019, https://ryanholiday.net/books-courses/

Kevin Hart, Entertainer

- "Sun doesn't stop for nobody, man. The sun doesn't stop. The sun is going to be up in the morning, regardless. Regardless of how I feel and how depressed I am, the sun is going to shine in the morning, and at nighttime, the moon is going to be there. And you're going to look up; these days are going to keep going by."

  So are you going to let the days go by and look up to find you've wasted a year doing... what? Or do you just pick it up, figure it out, admit you've made some mistakes, and move on? Life goes on. I'm a 'life goes on' type of guy." – Paraphrased for content. Source: Joe Rogan Podcast, April 6, 2019, https://www.youtube.com/watch?v=XW_KhFq4LQo

- "All storms eventually pass and when they do the sun shines bright as hell....Stay motivated & Stay focused....We are all great damn it!!!" Source: Kevin Hart, Twitter/X, May 20, 2020 https://twitter.com/KevinHart4real/status/1260242844226170880?s=20

Stephen Hawking, Physicist, Cosmologist and Author

- "Remember to look up at the stars and not down at your feet. Try to make sense of what you see and wonder about what makes the universe exist. Be curious. And however difficult life may seem, there is always something you can do and succeed at. It matters that you don't just give up." Source: The Guardian, 'Remember to look up at the stars', March 14, 2018 https://www.theguardian.com/science/2018/mar/14/best-stephen-hawking-quotes-quotations

- "The universe doesn't allow perfection." Source: A Brief History of Time, September 1, 1998 https://www.amazon.com/Brief-History-Time-Stephen-Hawking/dp/0553380168

Ryan Holiday, Best-Selling Author

- "Even during a quiet night at home, we're spending time thinking of our list of improvements we need to work on. There may be a beautiful sunset, but instead of taking it in, we're busy taking a picture of it. We are not present, and so we miss out. On life. On being our best." Source: Stillness is the Key, 2019, https://ryanholiday.net/books-courses/

Andrew Huberman PhD, Stanford University Neuroscientist

- "If you wake up before the sunrise, try to get some sunshine in your eyes once it comes out. If you miss that morning sun for more than a few days and you view evening sunlight you delay your circadian clock which makes you fall asleep later, wake later etc. Get morning light." Source: Andrew Huberman's Twitter/X, March 15, 2022, https://twitter.com/hubermanlab/status/1503813057012637703?lang=en

- "By looking at sunlight through a window, it's 50 times less effective than if that window were to be open — mostly because those windows filter out a lot of the wavelengths of blue light that are essential for stimulating the eyes and this wake-up signal." Source: Huberman Lab podcast, January 31, 2022 https://podcasts.apple.com/us/podcast/huberman-lab/id1545953110?i=1000549503046

- "The yellow-blue contrast present at sunrise (& orange-blue at sunset) = the optimal stimulus for setting your circadian rhythm for quality sleep and daytime mood, focus & alertness. Even w/cloud cover when it's not perceptible it's still there & is more effective than any artificial light." Source: Andrew Huberman's Twitter/X, September 3, 2023, https://twitter.com/hubermanlab/status/1698302038851903511

Dwayne Johnson, "The Rock"

- Reflecting on his sun and sunshine artwork as they represent: "the great struggle and overcoming that struggle and being appreciative of my success" Source: WWE, March 28, 2012 https://www.youtube.com/watch?v=DKYf1dVewXk

Abraham Lincoln, 16th US President

- "Behind the cloud the sun is still shining." Source: Illinois Landmark Plaque, https://readtheplaque.com/plaque/lincoln-s-farewell-to-illinois

Kelly McGonigal, PhD, TedTalk Speaker, Psychology Professor at Stanford University

- "For most of us, early morning is one time of the day we can be alone, without demands and distractions. Rising a bit early can allow you to experience inner stillness and offer your energy to a greater intention for your day." Source: Yoga Journal, March 2, 2010 https://www.yogajournal.com/yoga-101/shine

National Geographic

- Tides rise and fall in the ocean due to the gravitational pull of the sun and the moon. The moon has a bigger influence because it's closer, but the sun's gravity also affects tide changes. Source: National Geographic, Cause and Effect: Tides, https://education.nationalgeographic.org/resource/cause-effect-tides/

Daniel H. Pink, Author, Businessman

- "Greatness and nearsightedness are incompatible. Meaningful achievement depends on lifting one's sights and pushing toward the horizon." Source: Drive: The Surprising Truth About What Motivates Us, 2011

Naval Ravikant, Entrepreneur, Investor

- "We accept the voice in our head as the source of all truth. But all of it is malleable, and every day is new. Memory and identity are burdens from the past preventing us from living freely in the present." Source: Book, The Almanack of Naval Ravikant, 2020, https://www.navalmanack.com/

Tony Robbins, Entrepreneur

- "TODAY is a new day, a new beginning full of POSSIBILITY. Today is more precious than we can even comprehend!", Source: Tony Robbins Twitter/X April 13, 2023 https://twitter.com/TonyRobbins/status/1646509957200896007

Eleanor Roosevelt, Former First Lady

- "With the new day comes new strength and new thoughts." Source: My Day, January 8, 1936 newspaper column written by Eleanor, https://www2.gwu.edu/~erpapers/myday/displaydoc.cfm?_y=1936&_f=md054227

Franklin D. Roosevelt, 32nd President of the United States

- "We have always held to the hope, the belief, the conviction, that there is a better life, a better world, beyond the horizon." Source: American Presidency Project at UC Santa Barbara, October 12, 1940 https://www.presidency.ucsb.edu/documents/address-hemisphere-defense-dayton-ohio

Teddy Roosevelt, 26th President of the United States

- "Sunrise and sunset turned the sky to an unearthly flame of many colors above the vast water. It all seemed the embodiment of loneliness and wild majesty. Yet everywhere man was conquering the loneliness and wresting the majesty to his own uses." https://theodorerooseveltcenter.org/ Source: Through the Brazilian Wilderness, 1914.

Mother Teresa, Nun and Founder of the Missionaries of Charity

- "See the stars, the moon and the sun, how they move in silence… We need silence to be able to touch souls. Silence Gives Us A New Perspective." Source: Catholic Online and Catholic Company, https://www.catholic.org/clife/teresa/quotes.php and https://www.catholiccompany.com/mother-teresa-silence-gives-us-a-new-perspective-mug-i128335/

Eric Thomas, Author, Keynote Speaker, Coach

- "It doesn't matter whether you are a Lion or a Gazelle… when the sun comes up, you'd better be running." Source: Twitter/X, January 18th, 2012, https://twitter.com/Ericthomasbtc/status/159629105422270464

Neil deGrasse Tyson, Astrophysicist

- "How many earths would fit in the sun? About one million earths," Source: StarTalk, July 27, 2021, https://www.youtube.com/watch?v=apRVqNc4UhE
- "How far away is the sun in light travel time? it takes 500 seconds for the light to reach earth from the sun so 500 seconds if you do the math it's 8 minutes and 20 seconds." Source: StarTalk, July 27, 2021, https://www.youtube.com/watch?v=apRVqNc4UhE
- If the sun disappeared "from the center of our solar system we would continue to see the sun, [and] feel the sun until 8 minutes and 20 seconds passes." Source: StarTalk, July 27, 2021, https://www.youtube.com/watch?v=apRVqNc4UhE
- Due to refraction of sunlight, "When you see a sunrise, the sun is still below the horizon and has not yet risen. You're getting a few extra minutes of sun rise light." Source: StarTalk, July 28, 2020 https://www.youtube.com/watch?v=9bww_ux8NCo

Gary Vaynerchuk, "Gary V", Entrepreneur

- He prefers talking to a person, but when constructive feedback can only be written, he uses a Sun to show his feedback comes with warmth: "The misinterpretation of the written word digitally is a monster. People will consume it based on their framework. When I do [send a Slack] and I have to give any level of direct [feedback], I'm coming with a heart emoji and a sun right behind it." Source: The Verge, June 21,2023 https://www.theverge.com/23766700/gary-vee-vaynerchuk-vaynerx-vayner-media-influencer-content-marketing-social-media

Eddie Vedder, Musician

- "And sometimes is seen a strange spot in the sky. A human being that was given to fly… " Source: Pearl Jam, Given To Fly lyrics, 1998, https://pearljam.com/music/song/given-to-fly/lyrics

Venus Williams, Tennis Pro

- "If the sun comes up, I have a chance." Source: Guardian, June 25, 2012 https://www.theguardian.com/sport/2012/jun/25/venus-williams-wimbledon-2012

Jocko Willink, Ex-Navy SEAL, Author

- "The sunrise is better if you already paid for it." Source: Jocko Willink's Twitter/X January 22, 2019 https://twitter.com/jockowillink/status/1087739871534272517?lang=en
- "There will be dark times. But the sun will come up." Source: Jocko Willink's Twitter/X July 14, 2020 https://twitter.com/jockowillink/status/1283021438534860801
- "The task is simple: Do Work. Reward: Sunrise." Source: Jocko Willlink's Twitter/X, January 2, 2016, https://twitter.com/jockowillink/status/683301995864694784

www.ingramcontent.com/pod-product-compliance
Lightning Source LLC
Chambersburg PA
CBRC091205010526
44107CB00021B/1247